Nuclear Engineering Division 2021

Held at the 2021 AIChE Annual Meeting

Boston, Massachusetts, USA and Online
7 - 11 November and 15 - 19 November 2021

ISBN: 978-1-7138-5708-2

Nuclear Engineering Division 2021

Held at the 2021 AIChE Annual Meeting

Boston, Massachusetts, USA and Online
7 - 11 November and 15 - 19 November 2021

AIChE
The Global Home of Chemical Engineers

Printed from e-media with permission by:

Curran Associates, Inc.
57 Morehouse Lane
Red Hook, NY 12571

CURRAN ASSOCIATES INC. proceedings.com

Some format issues inherent in the e-media version may also appear in this print version.

AIChE
120 Wall Street, FL 23
New York, NY 10005-4020

Phone: (800) 242-4363
Fax: (203) 775-5177

www.aiche.org

Additional copies of this publication are available from:

Curran Associates, Inc.
57 Morehouse Lane
Red Hook, NY 12571 USA
Phone: 845-758-0400
Fax: 845-758-2633
Email: curran@proceedings.com
Web: www.proceedings.com

TABLE OF CONTENTS

632462 Molecular Dynamics Simulations of Hydrogen Retention in Single-Crystal and Polycrystalline Tungsten

Abstract:

When tungsten is exposed to hydrogen plasma, it is known to form blisters. The observation of blisters is difficult to explain in light of the fact that electronic structure calculations and molecular dynamics simulations show that hydrogen atoms embedded in tungsten repel each other. We study hydrogen blistering through molecular dynamics, and have verified the absence of hydrogen self-clustering, even at times on the order of hundreds of nanoseconds. Instead, hydrogen atoms are observed to push nearby hydrogen atoms out of the way, resulting in each hydrogen atom being effectively isolated from other hydrogen atoms. Nudged elastic band calculations of hydrogen entering a void show that as the number of hydrogens in the vacancy increases, the barrier becomes larger for any new hydrogen that could enter. Instead, we expect that grain boundaries may be needed for the formation of blisters. Our simulations of hydrogen near grain boundaries show that there is indeed a tendency for hydrogen to collect along grain boundaries, and that hydrogen does not leave the grain boundary plane or diffuse at significant rates once situated there. There is also a significant drift force that causes hydrogen to be attracted to grain boundaries, which we quantify. Finally, we investigate the mechanical properties of grain boundaries with hydrogen adsorbed on them to determine whether blisters might form by way of delamination of the grain boundaries under stress.

622258 Mechanical Properties of Plasma-Exposed Tungsten

Abstract:

Investigating the impact of helium (He) ion implantation on the mechanical properties of tungsten is of utmost importance for evaluating tungsten as a plasma-facing component (PFC) in nuclear fusion devices. Toward this end, here, we report results on the mechanical properties of single-crystalline tungsten containing structural defects, voids and over-pressurized He nanobubbles, related to the conditions of plasma exposure in nuclear fusion devices. Based on systematic molecular-dynamics computations using properly parametrized interatomic potentials, including recently developed machine learning potentials, we determine the elastic moduli and investigate the mechanical response of PFC tungsten beyond the elastic regime, computing mechanical properties such as tensile strength and characterizing in detail failure mechanisms upon introducing plasma-exposure-related defects into the tungsten structure.

Our simulations reveal that empty voids are centers of dilatation resulting in the development of tensile stress in the tungsten matrix, whereas He-filled voids (nanobubbles) introduce compressive stress in the plasma-exposed tungsten. We establish universal exponential scaling relations for the bulk, Young, and shear moduli, as well as the Poisson ratio of plasma-exposed tungsten as functions of its porosity. Furthermore, we find that the elastic moduli of plasma-exposed tungsten soften substantially as a function of He content in the tungsten matrix, also following exponential scaling relations, in addition to the softening caused with increasing temperature. A systematic characterization of the dependence of the elastic moduli on the He bubble size reveals that He bubble growth significantly affects both the bulk modulus and the Poisson ratio of plasma-exposed tungsten, while its effect on the Young and shear moduli of the plasma-exposed material is weak.

Furthermore, analysis of the mechanical behavior of PFC tungsten reveals that the presence of voids reduces the ultimate tensile strength (UTS) of tungsten, which is further affected by increasing its porosity. Introducing He into the voids leads to further reduction of the UTS with increasing He content. We find that the presence of He bubbles in the tungsten matrix causes embrittlement, mediated by crack initiation at the bubble/matrix interface; thus, He bubble formation and growth has a significant impact on the fracture mechanics of PFC tungsten. Our findings contribute directly to the further development of a structure–properties database that is required for the predictive modeling of the dynamical response of PFCs in nuclear fusion devices.

622051 Effects of Elastic Softening and Surface Hole Formation on Surface Morphological Evolution in Plasma-Facing Tungsten

Abstract:

Tungsten is the plasma-facing component (PFC) material for the divertor region of the International Thermonuclear Experimental Reactor (ITER) due to its exceptional thermomechanical properties, such as high melting point, mechanical strength, and thermal conductivity, as well as its low sputtering yield and low tritium retention. A large body of experimental evidence has established that PFC tungsten suffers severe surface degradation as a result of exposure to high fluxes of helium (He) and extreme heat loads. Specifically, experimental studies have shown that, under typical operating conditions expected in ITER, He implantation above a threshold of incident ion energy of approximately 35 eV causes formation of a fuzz-like nanostructure on the tungsten surface over the temperature range from 900 K to 2000 K. The formation of such surface nanostructure has adverse effects on the mechanical behavior and structural response of PFC tungsten as well as on the reactor performance.

Here, we report results on the surface morphological evolution of PFC tungsten and examine a number of factors that impact such evolution. Our analysis is based on self-consistent dynamical simulations according to an atomistically-informed, continuum-scale surface evolution model that has been developed following a hierarchical multiscale modeling strategy and can access the spatiotemporal scales of relevance to fuzz formation. The model accounts for PFC surface diffusion driven by the compressive stress originating from the over-pressurized helium bubbles in a thin nanobubble region, which forms in the near-surface region of PFC tungsten as a result of He implantation, in conjunction with formation of self-interstitial atoms in tungsten that diffuse toward the surface. The model also accounts for the softening of the elastic moduli of PFC tungsten, both thermal softening at high temperature and softening due to He accumulation in tungsten upon implantation, and the elastic moduli are time dependent and evolve until the He content in tungsten reaches its steady state. The dependence of the elastic moduli on the He content follows an exponential scaling relation predicted by molecular-dynamics simulations, while the He content in the near-surface region of PFC tungsten evolves according to a first-order saturation kinetics, consistent with experimental and simulation results reported in the literature. Our analysis establishes that this elastic softening accelerates nanotendril growth on the PFC surface and the onset of fuzz formation. We also introduce the concept of an incubation time as a kinetic metric for nanotendril growth on the PFC surface, which is equivalent to that of incubation fluence discussed in the literature, in order to predict and explain the minimum exposure time required to observe fuzz formation on PFC tungsten surfaces.

Furthermore, our simulations account, in an empirical fashion, for two types of subsurface bubble dynamical phenomena in the nanobubble region of PFC tungsten during He plasma irradiation, involving bubble bursting and surface crater formation. We demonstrate that the bubble-bursting-mediated surface hole formation effect on the PFC tungsten surface further accelerates the growth rate of nanotendrils and the onset of fuzz formation. As a result, the predicted incubation time for surface nanotendril growth is shortened, in agreement with experimental data of incubation fluence at comparable plasma exposure conditions. We also explore systematically the dependence of the PFC surface morphological response on the areal density of holes introduced at regular time intervals onto the He-implanted tungsten surface, a parameter in our analysis that serves as a proxy for the rate of He bubble bursting. More importantly, our simulations capture fine surface features in the PFC tungsten surface morphology and predict that the average spacing between nanotendrils is on the order of 100 nanometers, consistent with experimental findings.

630799 Helium Aggregation Near Grain Boundaries in Plasma-Facing Tungsten

Abstract:

Helium is known to accumulate in plasma-facing materials, forming bubbles. Helium transport is also slowed markedly by the presence of grain boundaries. With this in mind, we report the results of molecular dynamics simulations of helium plasma exposure for intervals of 0.5–1 μs of several tungsten bicrystals with low-Miller-index surfaces exposed to the plasma. These bicrystals have grain boundaries either intersecting the surface at right angles or parallel to the original surface approximately 6 nm below it. As expected, helium accumulates on the grain boundaries and is trapped there, but we also observe a "depleted region" around the grain boundaries up to 3.5 nm away from the grain boundary plane that is largely devoid of helium bubbles over this time interval. The distributions of helium atoms and clusters in the directions parallel to the grain boundary plane are largely indistinguishable from those from single-crystal simulations—it is only the out-of-plane distribution that is significantly different. The presence of grain boundaries does not seem to impede the formation of adatom "islands" generated by prismatic dislocation loops that are annihilated on the surface, but some surface orientations allow the dislocations to traverse the grain boundary plane rather than escaping into the bulk, creating a ridge at the junction of the grain boundary with the surface. Grain boundaries parallel to the surface essentially act as a barrier for helium, trapping any helium atoms and preventing them from making their way deeper into the crystal. This results in a very abrupt change in the helium distribution as a function of depth at the grain boundary, with only a few stray atoms making their way into the grain facing away from the plasma.

623157 Soret Diffusion of Helium and Intrinsic Point Defects in Tungsten

Abstract:

Plasma-facing materials (PFMs) in nuclear fusion reactors must withstand extreme conditions, with high heat and particle fluxes that modify their microstructure. These fluxes create strong temperature gradients and highly non-equilibrium concentrations of diverse species in the PFM. In addition to the helium (He) ash and hydrogenic species, neutrons generated in the fusion reaction will create intrinsic point defects, namely, vacancies and self-interstitial atoms (SIAs), and their clusters. Additional vacancies and SIAs are generated by trap mutation reactions undergone by helium clusters in the PFM bulk and in regions near the plasma-exposed surfaces. These defects, as well as He atoms and small mobile helium clusters, will then migrate in the presence of concentration and temperature gradients.

In this work, we use non-equilibrium molecular-dynamics (NEMD) simulations to study the transport of He and SIAs in the presence of a thermal gradient in tungsten PFM and discover evidence for Soret diffusion of He and SIAs in tungsten. We find that, in all cases, the intrinsic point defects and impurity atoms tend to migrate toward the hot regions of the material and calculate their concentration profiles in the direction of the temperature gradient. We also analyze thermal and species transport in tungsten within the framework of irreversible thermodynamics, which has been implemented in the cluster-dynamics code Xolotl. The resulting concentration profiles from the NEMD simulations are in agreement with the predictions of irreversible thermodynamics. We compute a negative heat of transport for each species analyzed, which indicates that the respective driven species fluxes are directed opposite to the heat flux. We demonstrate that when Soret diffusion or thermomigration, *i.e.*, drift species transport driven by the thermal gradient, is considered, the resulting steady-state profiles differ significantly from those when species transport is decoupled from heat transport.

632136 A Revision of Classical Force Fields for Tri-*N*-Butil Phosphate Molecular Dynamics Simulations

Abstract:

Tri-*n*-butil phosphate (TBP) is a solvent extraction ligand commonly used in the separation of metals from acidic aqueous solutions in nuclear chemical engineering. It's science and technology has been reported in a well-known 4-volume book series in the mid 1980's but advances in computing had not been used to shed light at atomistic level phenomena at that point in time. Starting in the mid 1990's, and continuing to 2010, molecular dynamics simulations were used to provide atomistic insight on solvent extraction. In the past decade a number of articles appeared in the literature employing classical molecular dynamics for studying various aspects of chemical complexation in solvent extraction involving TBP. It is instructive now, in view of growing computing power and access to open-source scientific software, for a computational revision of proposed force fields to be carried out with the intent of further advance the science and technology of TBP in more realistic scenarios where competing extraction of acid, water, and additional metals take place.

This work is reproducing and revising significant classical force fields published in the open literature in accord with current open science standards to compute thermochemical properties of TBP from atomistic simulations. It is expected that this work will provide recommendations on force fields for TBP to be used in additional studies. Our own past work has revelead an important effect of the electric dipole moment of TBP in modifying the interfacial property of aqueous/organic interfaces that leads to the extraction of water. We expect a similar effect on the extraction of other solutes therefore additional research in this direction is warranted. Past work of interest and basis for this project:

Ye, X., Cui, S., de Almeida, V. F., Khomami, B., 2009 Interfacial complex formation in uranyl extraction by tributyl phosphate in dodecane diluent: a molecular dynamics study, J. Phys. Chem. B, 113, pp. 9852–623

Ye, X., Cui, S., de Almeida, V. F., Hay, B. P., Khomami, B., 2010, Uranyl nitrate complex extraction into TBP/dodecane organic solutions: a molecular dynamics study Phys. Chem. Chem. Phys. 12, pp 15406–94

Ye, X., Cui, S., de Almeida, V. F., Khomami, B. 2013 Molecular Simulation of Water Extraction into a Tri-n-Butylphosphate/n-Dodecane Solution J. Phys. Chem. B 117 pp. 14835–415

Cui, S., de Almeida, V. F., Hay, B. P., Ye, Xianggui, Khomami, B. 2012 Molecular Dynamics Simulation of Tri-n-Butyl-Phosphate Liquid: A Force Field Comparative Study J. Phys. Chem. B 116 pp. 305–136

Ye, X., Cui, S., de Almeida, V. F., Khomami, B. 2013 Effect of Varying the 1–4 Intramolecular Scaling Factor in Atomistic Simulations of Long-Chain N-Alkanes with the OPLS-AA Model J. Mol. Model. 19 pp 1251–58

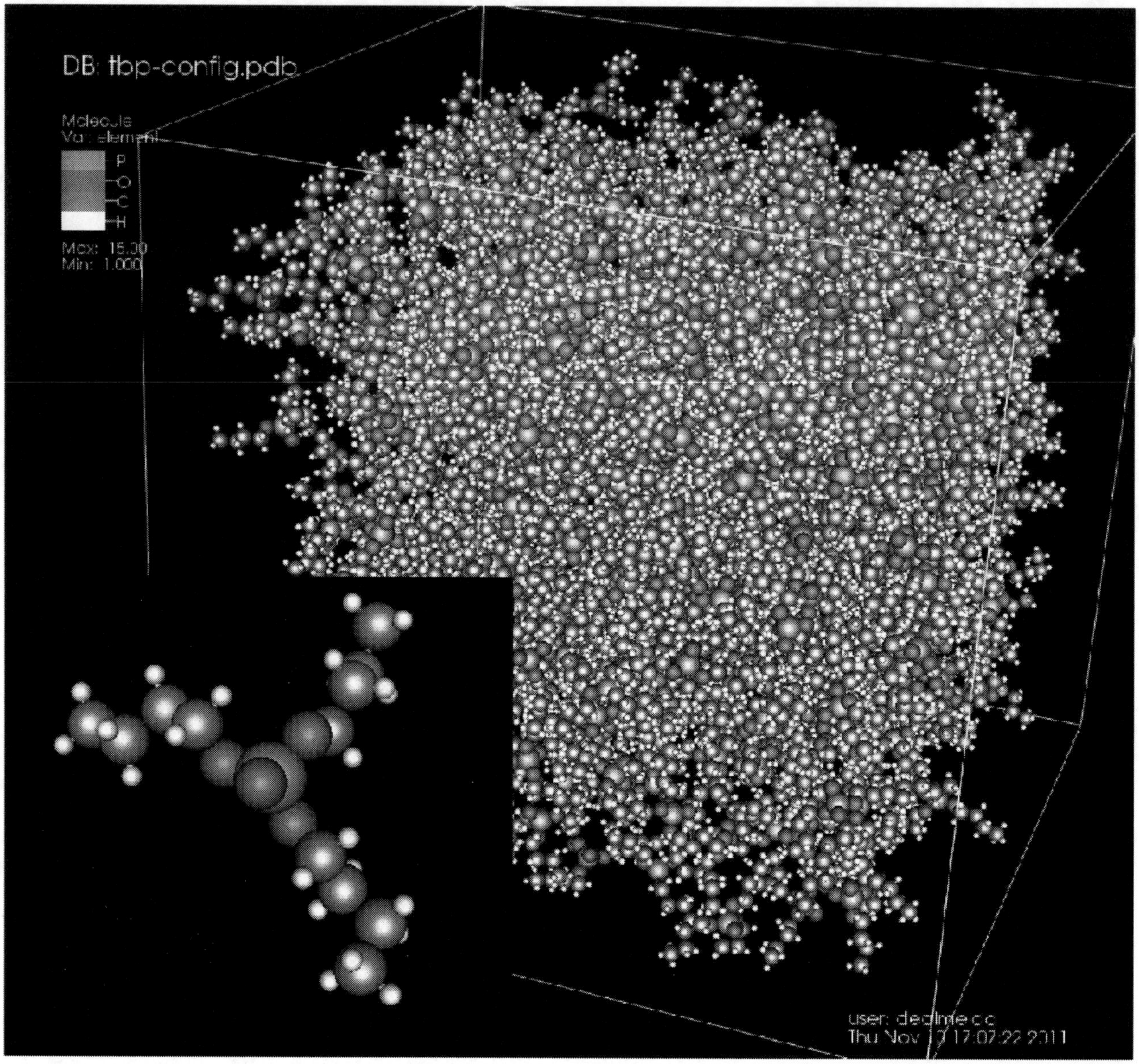
DB: tbp-config.pdb
Molecule
P
O
C
H
Max: 15.00
Min: 1.000

622922 On the Onset of 'Fuzz' Formation in Plasma-Facing Materials: A Hierarchical Multiscale Modeling Approach

Abstract:

Fusion energy is the most promising carbon-free and sustainable energy source. However, harnessing fusion energy for the electrical grid is extremely challenging. One major problem that makes the realization of fusion energy particularly challenging is the severe surface degradation of plasma-facing-component (PFC) tungsten in linear plasma devices and fusion tokamaks. Tungsten is the PFC material for ITER (International Thermonuclear Experimental Reactor) and exhibits formation of nanometer-sized tendrils, or porous "fuzz", following low-energy helium exposure. Helium (He) produced in the deuterium-tritium fusion reaction is a natural impurity in the plasma and it is extracted from the plasma through the divertor in ITER. Experimental studies have revealed that high particle fluxes of this low-energy helium produce fuzz-like nanofibers at the tungsten divertor within the temperature range from 900 K to 2000 K (consistent with the temperatures expected at the divertor wall due to the anticipated heat loads), which has adverse effects on the reactor performance.

Fuzz formation in plasma-facing tungsten is a complex multi-physics phenomenon resulting from the combined effects of driven surface diffusion, subsurface bubble dynamics, bubble bursting and surface crater formation, dislocation loop punching, anisotropies in material properties, changes in material thermophysical properties in the damaged tungsten, and many more. Developing a predictive model capable of capturing the formation and evolution of the fuzz-like complex surface morphology mediated by the underlying dynamical processes that are characterized by disparate spatiotemporal scales has been a major theoretical and computational challenge. While large-scale molecular-dynamics (MD) simulations have been used to successfully identify many of the aforementioned phenomena [1], simulating experimentally the spatiotemporal scales (mm-hr) of relevance to fuzz formation is well beyond the scope of such atomistic simulations. Therefore, we followed a hierarchical multiscale modeling paradigm to develop a continuum-scale model for simulating the onset of fuzz formation in He plasma-irradiated tungsten [2].

In our modeling framework, large-scale MD simulation results are used to parameterize the constitutive equations required for the closure of the continuum-scale model and targeted MD simulations are used to determine the thermophysical properties [3] of He-implanted tungsten. One highlight of this framework is the calculation of the bubble-matrix (He-W) interfacial free energy, a key parameter in the equation of state (EOS) for the He density in the bubble [4]. We have employed a novel iterative scheme, employing a protocol of MD simulations and the Young-Laplace equation, to solve the nonlinear EOS and determine the value of the bubble-matrix interfacial free energy. Furthermore, we have used a cluster-dynamics model to simulate the subsurface helium bubble evolution and bubble bursting and develop a coarse-grained model for He concentration evolution in plasma-facing tungsten, which can be coupled with the surface evolution model. Our cluster-dynamics model, implemented numerically in our code Xolotl, has been parameterized based on large-scale MD simulation results and benchmarked against experimental measurements [5]. Based on this atomistically-informed continuum-scale model, we have conducted self-consistent numerical simulations of the He-implanted tungsten surface morphological evolution and validated the model by comparing the simulation predictions with experimental measurements.

[1] K. D. Hammond, S. Blondel, L. Hu, D. Maroudas, and B. D. Wirth, Acta Mater. **144**, 561-578 (2018).

[2] D. Dasgupta, R. D. Kolasinski, R. W. Friddle, L. Du, D. Maroudas, and B. D. Wirth, Nucl. Fusion **59**, 086057 (2019).

[3] A. Weerasinghe, B. D. Wirth, and D. Maroudas, ACS Appl. Mater. Interfaces **12**, 22287–22297 (2020).

[4] K. D. Hammond, D. Maroudas, and B. D. Wirth, Sci. Rep. **10**, 2192 (2020).

[5] S. Blondel, D. E. Bernholdt, K. D. Hammond, and B. D. Wirth, Nucl. Fusion **58**, 126034 (2018).

638634 Reprocessing of Highly Enriched Aluminum-clad Fuels in the Savannah River Site H-Canyon Facility: Past, Present, and Future

Abstract:

Aluminum-clad spent nuclear fuel was used in both retired nuclear weapons program production reactors and in research reactors. Aluminum-clad fuel is intended to be reprocessed – not stored for long periods to time – due to its reduced robustness compared to other cladded fuels. At present, the H-Canyon on the Savannah River Site is the only remaining operational nuclear chemical separations facility in the United States capable of largescale nuclear material processing; aluminum-clad fuels have been and are being reprocessed in the H-Canyon facility to reduce the existing fuel stockpile. This presentation will cover the unique nuclear material processing that is performed in the H-Canyon facility and the presenter's own research work that has supported it over the past several decades.

620468 Hanford Hlw Sludge Processing in Crossflow Filtration

Abstract:

The Pretreatment Facility is the first step in the process of vitrifying Hanford's tank waste. Filtration, washing and/or caustic leaching to pretreat the Hanford sludge is required before it undergoes high-level waste (HLW) vitrification. Samples of actual Hanford tank waste sludge were evaluated by the pretreatment processes of crossflow filtration, washing, caustic leaching, and rinsing. Filtration of the waste is necessary to remove components such as aluminum, chromium, sodium, and phosphorus that are soluble in water and often limit the waste loading in the glass. Data collected at 5 M Na and 20 °C demonstrated flux increases dramatically with washing. Additionally, relationships between Na molarity, viscosity, and solids loading impacts on filter performance were used to develop an empirical fit that represents real waste testing as a function of the gel concentration

630766 Application of Blind Source Separation and Partial Least-Squares Regression to Quantify Target Species in Complex Nuclear Waste Mixtures

Abstract:

The proposed treatment of waste stored at the Hanford site in Washington State is expected to continue for many years. One of the main challenges is the complexity of the waste, which comprises three phases, 25 chemical constituents and 46 radionuclides. [1] Extensive sampling and analysis is to take place during processing. Online monitoring will be used to supplement and replace off-line sample collection, which will reduce worker exposure to hazards of handling the waste and improve the operation of the process. [2]

As part of our research, process analytical technology (PAT), such as *in situ* Raman and infrared spectroscopy, have been developed to measure the waste composition in real time. In this work we present a flexible data-driven modeling framework, in which we develop a signal-to-composition model based on knowledge of the liquid low-activity waste (LAW) system. The LAW contains many species, out of which nitrate, nitrite and sulfate need to be quantified in real time to facilitate continuous processing. In the usual approaches, a chemometrics model, such as partial least-squares regression (PLSR) is developed and trained with all the species in the waste. However, such an approach requires extensive model calibration involving both the target species (such as nitrate, nitrite and sulfate), and the many other species that may not need to be quantified. In order to reduce the number of experiments required to build a PLSR model, we propose a new methodology that couples PLSR with blind source separation (BSS). BSS methods, such as independent component analysis (ICA), use little or no prior information to decompose mixture data into its constituents, relying on minimization of the mutual information or maximization of the non-Gaussianity of the data. [3]

In our work, BSS is used as a preprocessing step, to identify and remove the signals of the non-target species. The blind source separation (BSS) – partial least-squares regression (PLSR) is carried out in four steps:

- Apply BSS methods, such as independent component analysis (ICA) and multiple curve resolution – alternating least-squares (MCR-ALS), which use the mixture spectra to identify a sources and mixing matrices, to identify the independent components (sources) in the waste mixtures;
- Compare the independent component (sources) spectra to a library of reference spectra of the target species using a matrix of correlation coefficients. Based on the correlation, classify the independent sources in two categories: target species (which correlate with the library) and non-target species (which do not correlate with the library);
- Calculate and remove the signals of the non-target species from the mixture spectra;
- Develop a partial least-squares regression model to quantify the targets. Since the signals of the non-target species have been removed, the model can be trained with only the targets, reducing the need for extensive calibration.

The BSS-PLSR framework was tested on Raman and infrared spectroscopy data using simulated and experimental measurements. Preprocessing the data with BSS resulted in greater accuracy in the concentration predictions for both spectroscopic techniques. One of the main advantages of using BSS to preprocess the data is removing the dependence of the PLSR model on the non-target species.

Therefore, even if the number and concentrations of non-target species fluctuates throughout the process, the training data set for the PLSR does not need to be updated. While the BSS-PLSR framework was tested on Raman and IR spectra, it can theoretically be applied to any other spectroscopic technique.

[1] Peterson, R. A.; Buck, E. C.; Chun, J.; Daniel, R. C.; Herting, D. L.; Ilton, E. S.; Lumetta, G. J.; Clark, S. B. Review of the Scientific Understanding of Radioactive Waste at the U.S. DOE Hanford Site. *Environmental Science & Technology* **2018**, 52, 108381–396

[2] Poirier MR, Howe AM, Miera FR, Stone ME, DiPrete CC, Farrar ME. WTP Real-Time In-Line Monitoring Program Tasks 4 and 6: Data Quality and Management and Preliminary Analysis Plan. **2017**.

[3] Hyvärinen, A. & Oja, E. Independent component analysis: algorithms and applications. *Neural Networks* **13**, 411–430 (2000).

627002 Adsorption of Organic Iodides from Vessel Off-Gas (VOG) Streams on Silver-Containing Adsorbents

Abstract:

During the nuclear fission process, radioactive organic iodides are generated in the gas form including CH_3I, C_2H_5I, ..., $C_{12}H_{25}I$. For environmental protection purpose, silver-containing materials such as reduced silver exchanged mordenite (Ag^0Z), silver functionalized silica aerogel (Ag^0-Aerogel) and silver nitrate impregnated alumina (AgA) are used to adsorb these contaminants in the vessel off-gas stream (VOG). In the presented work, the organic iodide (CH_3I) adsorption at extremely low concentrations (ppb levels) was studied.

By performing the adsorption of 104 and 1044 ppbv CH_3I at 150 °C on Ag^0Z, Ag^0-Aerogel and AgA, the Ag^0-Aerogel has been identified as the most effective adsorbent for its high adsorption rate and strong aging resistance. Therefore, further studies including the temperature effect and alkyl group effect, were conducted using Ag^0-Aerogel.

The 104, and 1044 ppbv CH_3I adsorption experiments at 100, 150 and 200 °C on Ag^0-Aerogel were conducted, and an abnormal increase of uptake rate was observed at 104 ppb, 200 °C (200% - 300% higher than that at 150 °C). To explain this behavior, firstly, a step-wise potential reaction pathway was proposed. The model includes the cleavage of CH_3I, binding of the free radicals, and the diffusion of the gas form product. Since diffusion limitation of the product is considered in this pathway, it can be further used to explain the abnormal behavior. Secondly, the porosities of the Ag^0-Aerogel at different drying conditions (100, 150 and 200 °C) were measured using the nitrogen titration method. The results indicate that as the temperature decreases, the increasing in-pellet water concentration decreases the pore surface area and may decrease the availability of the silver sites during the adsorption. Thirdly, physical analyses including scanning electron microscopy with energy dispersive x-ray analysis (SEM-EDX) and x-ray photoelectron spectroscopy (XPS) were performed. The spectra show that an additional Ag-I compound is formed at 104 ppb, 200 °C, which may also lead to the significant increase in the uptake rate.

After reviewing the potential explanations, the significances of the factors influencing the uptake rate appear to be (from the most to the least), formation of an additional Ag-I compound, temperature effect introduced by the Arrhenius relationship, the decrease of the diffusion limitation and the increase of the silver site availability.

623798 Implementation of a New Defoamer in the Defense Waste Processing Facility to Reduce Batch Processing Time and Improve Safety

Abstract:

The legacy of the waste from the production of nuclear weapons materials is the biggest environmental challenge in the state of South Carolina, and one of the biggest in the United States. Although great progress has been made in processing and vitrifying high level waste at the Savannah River Site, approximately 35 million gallons remains to be processed. In this paper, we discuss the new defoamer that has been developed for use in processing high level radioactive waste. The use of the new defoamer will allow higher evaporation rates, leading to as much as a 50% decrease in processing time per batch.

The defoamer is novel because it is effective in the harsh conditions needed for processing high-level radioactive waste, including high temperatures, a wide pH range (4 to 13), strong oxidizing and reducing agents, and the presence of a myriad of metal catalysts. The defoamer has replaced an old antifoam agent that was chemically unstable at non-neutral pH and decomposed to form three flammable gases. Because the old antifoam agent decomposed quickly, processing changes were needed which resulted in increasing processing time, including lowering acid addition rates and evaporation rates.

The new defoamer is not only more effective at controlling foam but, as a result, will shorten the batch processing time and lead to safer and more efficient processing by eliminating the production of flammable gases. Testing demonstrated that the new defoamer would meet all the requirements needed for foam mitigation during DWPF processing. The new defoamer will shorten the mission lifetime, allowing the vitrification facility to be shut down several years early, potentially saving the federal government hundreds of millions of dollars in operating costs.

This paper will discuss the implementation of the new defoamer in the Defense Waste Processing Facility (DWPF). Use of the new defoamer has demonstrated that excessive foaming and foamovers have been eliminated allowing DWPF to increase acid addition rate and steam flow rate to maximum rates permitted by procedures and safety requirements.

633803 Water Sorption/Desorption Characteristics in Electrorefiner Salt-Occluded Zeolites

Abstract:

Electrorefiners currently operating in the Fuel Conditioning Facility of Idaho National Laboratory contain radioactive, actinide-containing molten chloride salt. The primary components of this salt are LiCl and KCl in a eutectic composition. The salt also contains UCl_3, $PuCl_3$, and numerous fission products from processing irradiated sodium-bonded metal fuel from the Experimental Breeder Reactor-II. It has been proposed to occlude the salt into a zeolite such as zeolite-4A or Na-Y zeolite before eventual conversion into ceramic waste forms for permanent disposal. In this work, we studied the characteristics of the salt-occluded zeolite for hydration and if such hydration is reversible. Zeolite-4A with 20 wt% LiCl-KCl was found to absorb up to about 10 wt% water when in contact with 40°C air at 20% relative humidity. As is the case with zeolite-4A without any occluded salt, essentially all of the moisture can be desorbed by heating the salt occluded zeolite-4A to 500°C under flowing argon. However, the dehydration is accompanied by a small amount of HCl gas formation due to hydrolysis of the salt. HCl generation was measured continuously while the salt occluded zeolite was heated to 500°C using an off-gas sparged auto-titrator. The amount of HCl generated ranged from about 0.3 to 0.5% of the total Cl contained in the salt for any given sample of salt occluded zeolite. The effect of temperature ramp rate and maximum temperature on the amount and rate of HCl generated was studied and will be reported.

631169 Performance Assessment for the Low Activity Waste and Intermediate Level Vaults (LAWV and ILV) at the Savannah River Site

Abstract:

Low Activity Waste (LAW) generated at the Savannah River Site (SRS) that exceeds disposal limits for burial in ground trenches is being disposed of in the LAW Vault (LAWV) located in E-Area. The LAWV is a large, above-grade, reinforced concrete structure 643 ft long, 145 ft wide, and 27 ft high at the roof crest. It is divided into 12 cells designed to contain more than 12,000 B-25 boxes. Intermediate level waste generated at SRS is being disposed of in the Intermediate Level Vault (ILV) also located in E-Area. The ILV is a largely below-grade, reinforced concrete vault with a footprint of 279 ft by 48 ft and consists of two modules: the Intermediate Level Tritium (ILT) module, which contains two 26-ft deep cells, and the Intermediate Level Non-Tritium (ILNT) module, which contains seven 28.4-ft deep cells. At the end of radwaste-generating operations at SRS, both LAWV and ILV will be closed, sealed, and covered with an earthen closure cap for long-term disposition.

Groundwater models have been developed for the ILV and LAWV using the PORFLOW software package with the purpose of evaluating dose to members of the public following site closure and thereby producing waste disposal limits. Modeling results are presented and discussed in this paper.

632329 Performance Assessment for Naval Reactor Component Disposal Areas (NRCDA) at the Savannah River Site

Abstract:

Naval Reactor (NR) waste generated by the reactor components from the U.S. Navy is comprised of highly radioactive components consisting of activated corrosion-resistant metal alloy contained within welded thick-walled steel casks, and auxiliary equipment primarily contaminated with lower levels of Activated Corrosion Products residing on the metal surfaces and contained within thinner-walled bolted containers. The NR waste has been stored and disposed in the two Naval Reactor Component Disposal Areas (NRCDAs), 643-7E and 643-26E, at the Savannah River Site (SRS).

Groundwater models have been developed using the PORFLOW software package for the two NRCDAs with the purpose of evaluating dose impacts and producing disposal limits for the SRS E-Area Low-Level Facility (LLWF). Modeling results are presented and discussed in this paper.

627761 Continuous-Flow Centrifugal Solid-Liquid Separation for the Recovery of Rare-Earth Elements Containing Particles from Phosphoric Acid Sludge

Abstract:

Phosphoric acid sludge contains concentrated phosphoric acid (up to 54% P_2O_5) and solid precipitates, including rare earth elements (REEs) at concentrations near 2,200 ppm. Low cost recovery of valuable P_2O_5 and simultaneous separation of solid particles from the phosphoric acid sludge could be an economically feasible approach to recovering REEs, while increasing the production of phosphoric acid. The phosphoric acid sludge, however, is a complicated fluid stream that cannot be simply separated by traditional separation technologies such as batch centrifugation, decantation, settling, or filtration because of its high viscosity and high solids content (e.g., 30–40%). Guided by a force balance model, a highly efficient solid/liquid separation method was demonstrated in this study, using a continuous-flow centrifugal contactor. A high shear is typically used to increase the contact surface between two liquid phases, thus enhancing the interphase mass transfer rate, and a centrifugal force is exerted on the fluids, inducing phase separation at the exit. By introducing the sludge directly through the phase separation region, we demonstrated sludge recovery and studied the effects of process parameters on solid particle capture with the goal to optimize the separation process. Three contactors in series yielded 94% solid recovery and 98% recovery of the liquid component. After saturation, the slurry was withdrawn and the accumulated solids in the contactor were flushed out. Recovered solid particles contained 1,500–2,895 ppm of REEs, which can be recovered after further processing.

Effective Adsorption of Chromium from Tannery Wastewater Using Green Synthesis Nano-Zero Valent Iron (GT-nZVI)

Ahmed S. Mahmoud[1], M. S. Mahmoud[2], Ahmed M. Noureldin[2], Robert W. Peters[3], and Mohamed K. Mostafa[4]

[1]*Scientific Research Development unit, Egyptian Russian University (ERU), Badr, Egypt*

[2] *Sanitary and Environmental Institute (SEI), Housing and Building National Research Center (HBRC), Giza, Egypt.*

[3]*Department of Civil, Construction and Environmental Engineering, University of Alabama at Birmingham, Birmingham, AL, USA*

[4]*Faculty of Engineering and Technology, Badr University in Cairo (BUC), Cairo, Egypt*

Abstract

Nowadays, Egypt built the biggest industrial area for tanning industries; this industry produces a vast amount of chromium in the tannery wastewater effluent. This study successfully prepared and characterized GT-nZVI by reducing agents extracted from soft black tea. The prepared GT-nZVI was characterized using x-ray diffraction (XRD), scanning electron microscopy (SEM), and energy dispersive analysis of x-rays (EDAX) analysis. The chromium removal efficiency reached 97.1% and 73.7% from an initial chromium concentration of 1.0 and 9.0 mg/L, respectively. The optimum removal efficiencies were achieved using 0.6 g/L GT-nZVI at acidic pH 5 after 30 min of shaking at a rate of 250 rpm. The equilibrium isotherm and kinetic results indicated that the adsorption of chromium onto GT-nZVI agrees with the Koble-Corrigan model with the lowest summation of errors of 0.394 and a Pseudo-Second-Order adsorption with the lowest summation of errors of 0.07. An artificial neural network (ANNs) was built to describe the factors affecting the adsorption process by 6 inputs, 4 hidden layers, and 1 output. The results obtained indicated that the initial concentration and GT-nZVI dose are the most significant variables that influence the adsorption process. The response surface methodology (RSM) equation was developed to allow the prediction of chromium removal efficiency at different operating parameters. Finally, this study strongly supports using GT-nZVI for chromium removal from tannery effluents.

1. Introduction

Water is the most traditional polar solvent with the capacity to exceed 75% of total earth capacity (Weingärtner and Franck, 2005, Mahmoud et al., 2021b, Farag et al., 2018). It is responsible for humans, plant, and animals nutrition and it is necessary for development especially in developing countries so, it considers as a principal source of life (Pahl-Wostl et al., 2008, Mostafa and Peters, 2015). Because of water polarity, it has multiple usages and can dissolve most of the Inorganic and Organic compounds and cause water and wastewater contamination (Schwarzenbach et al., 2010, Jason A. Heberling, 2018). Increasing water consumption, by increasing world population according to the estimates of the United Nations Organization the world's population October 2011exceeding than seven billion, increases global awareness about it to save water by decreasing consumption and water recycling (Dudgeon et al., 2006, Kennedy, 2011, Carpenter et al., 2011). Recently, water is becoming scarcer globally with evidence that it will become rarer. The water scarcity problem is the main reason for many wars in that world, especially in the Middle East (Beaumont, 1994, Dolatyar and Gray, 2000, Amery, 2002, Barnaby, 2009, Hamdy et al., 2019). There are different types of water resources include surface water such as rivers, seas and oceans, and groundwater (Mostafa and Peters, 2016a, Mostafa and Peters, 2016b). About 44% of the world population using groundwater as the second source of drinking water, such as China, the United States of America (USA), and Egypt (Shiklomanov, 1991, Pimentel et al., 2004, Taylor et al., 2013). In China, more than one-third of total water is demanded and some people using groundwater as the main source of drinking water (Oki and Kanae, 2006, Mostafa et al., 2017, SaryEl-deen et al., 2017). In the United States, Nearly 50% of Californian's people using groundwater as the main source of drinking water and agriculture and extract more than 14 billion gallons per day so it required to organize from scientists and government to overcome groundwater contaminations(Barlow and Clarke, 2017, Mostafa and Peters, 2017).

Wastewater is the most observable environmental problem among many in urban areas (Ngoc and Schnitzer, 2009, El-Shafei et al., 2016). The increasing world population increases water consumption and produces a huge amount of wastewater every day. There is an expectation of increasing the demand for sanitation by 2 billion will live in towns and cities, mainly in developing countries within 20 years (Langergraber and Muellegger, 2005). More than 90% of developing country's sewage is discharged untreated and polluting lakes, rivers, and coastal areas (Islam and Tanaka, 2004, Liu and Diamond, 2005). Conventional sanitation techniques, based on water-wasting technology, flush toilets, aren't a suitable solution for industrial and developing countries (Weng, 2002, Del Porto, 2005, Diaz and Barkdoll, 2006). The sanitation systems were built and designed on the premise that human excreta are a waste suitable for disposal into the environment and it can assimilate this waste (Esrey and Andersson, 1999, Langergraber and Muellegger, 2005).

The tanning industry is one of the most important industries all over the world. Throughout ancient history, the Ancient Egyptians employed the tanning process to cover the local needs of leather, belts, footwear, and drums (Forbes, 1956, Mahmoud et al., 2021b). By increasing the world population, increased the requirement of different leather products, leading to creating large tanneries projects such as Al Robaiky region, Egypt (Frosch and Gallopoulos, 1989). In Egypt, tanning businesses are the most significant source of national income (Nelson and Rosenberg, 1993, M Abdeldayem et al., 2020). The total number of tannery projects in Egypt is about 300 and most of them using the chromium tanning technique (Hafez et al., 2002, Walsh and O'Halloran, 1996). The tanning process using the chromium salts technique is effective, fast, stable, and low cost during the operation and production processes (Gupta et al., 2001, Sreeram and Ramasami, 2003). More than 65% of used chromium salts react with leathers and the chromium leftover remains in the effluent (Karabay, 2008, Kurniawan et al., 2011). So, tannery wastewater is recognized as highly contaminated industrial effluent including heavy metals, toxic chemicals, chloride, lime with high dissolved and suspended salts, and other pollutants (Chowdhury et al., 2013).

Several treatment technologies, like filtration, electrocoagulation, chemical precipitation, ion exchange, bioremediation, and adsorption, were tested for chromium removal from aqueous solutions and real wastewater ((Morsy et al., 2020)El-Shafei et al., 2016). Additionally, many papers have been published addressing the utilization of nano materials, especially iron nano particles, which is a promising material for the removal of heavy metals from wastewater due to its high reactivity and surface area (Li et al., 2020). This study attempts to use low-cost green synthesis of nano-Zero Valent Iron (GT-nZVI) sorbent material for chromum removal from aqueous solutions and real tannery wasteater. Batch adsorption experiments were conducted using GT-nZVI by adapting the operating conditions, viz., solution pH, GT-nZVI dosage, stirring rate , and adsorption time. Response surface methodology (RSM) and artificial neural network

(ANN) were also applied to find the correlation between the experimental factors and chromium removal efficiency

2. Material and Methods

2.1. Preparation of GT-nZVI

The extraction of reducing agents from soft black tea as well as reduction of metal cation from its soluble salt was adapted from previous studies (Mahmoud et al., 2019a, Mahmoud et al., 2020a).

2.2. Batch studies

When green synthesized nZVI (GT-nZVI) are placed in standard solutions of chromium (Cr) and agitated for an adequate time, the adsorption process occurs. Aqueous solutions (100 mL) were used and the adsorption capacity was determined at different environmental conditions (pH, Dose, Time, Stirring rates, and Concentrations). After filtration, the withdrawn samples were analyzed by atomic absorption spectroscopy according to the standard method for examination of water and wastewater 23 edition 2013 (Baird, 2017). Then the adsorption capacity and isothermal studies were conducted using adsorption isotherms (Mahmoud et al., 2020a, Mahmoud and Mahmoud, 2021, Rabie S. Farag and Peters, 2019). The percentage of removal efficiency was calculated using Eq. 1 (Mahmoud et al., 2021a, Rabie S. Farag and Peters, 2018). The amount of sorbed contaminants were calculated using Eq. 2 (Maha M. Elshfai and Elsaid, 2021, Mahmoud et al., 2021a, Rabie S. Farag and Peters, 2018). All experiments were performed in triplicate and all results were within lab measurements (Mahmoud et al., 2020b, Mahmoud et al., 2019b).

$$Sorption\ (\%) = \left(\frac{C_0 - C_e}{C_0}\right) \times 100 \quad (1)$$

where C_o is the initial concentration (mg/L) and C_e is the equilibrium concentration in solution (mg/L).

$$Q_e(mg/g) = \frac{(C_0 - C_e)V}{m} \quad (2)$$

where q_e is the equilibrium adsorption capacity (mg/g), V is the volume of aqueous solution (L), and m is the dry weight of the adsorbent (mg) (Mahmoud et al., 2018).

Table 1. Batch experiments at different operating parameters.

Effect	Dose (g)	Contact time (min)	pH	Stirring rate (rpm)	Initial concentrations (mg/L)
Effect of pH	0.6	30	3-11	250	5
Effect of adsorbent dose	0.1-1.0	30	5	250	5
Effect of contact time	0.6	10-50	5	250	5
Effect of stirring rate	0.6	30	5	100-300	5
Effect of concentration	0.6	30	5	250	1-9

2.3. Neural network structure

An artificial intelligence network using the backpropagation technique was familiar to approximation the adsorption of chrome onto GT-nZVI, the importance of each operating condition pH, dose, time, stirring rate, and concentrations. Also, to predict the corrected relation between predicted values and true measurements. The ANN model combined the input covariable (pH, dose, time, stirring rate, and concentrations) with hidden layers (weight and bias), and the final output removal layer. The estimated ANN structure for DY 50 removal using nano bentonite was expressed as 5 – 3 – 1 as shown in Figure 1 (Mahmoud et al., 2020a, Mahmoud et al., 2019a).

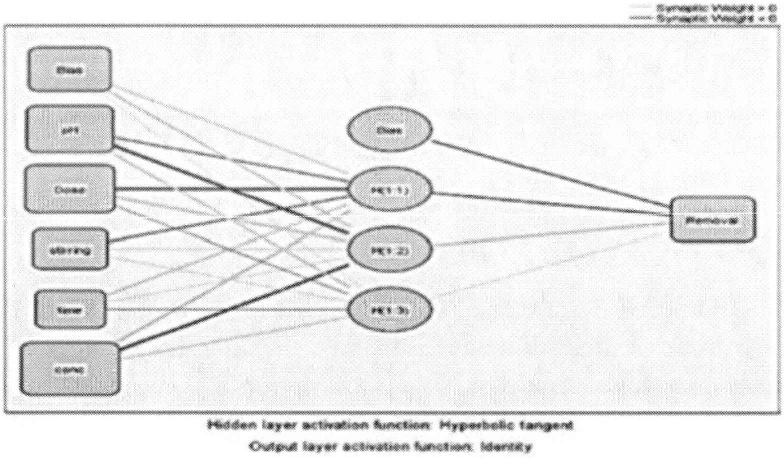

Figure 1. ANN of 5 – 3 – 1 structure used for the prediction of Cr removal efficiency.

2.4. Response surface methodology

A linear equation was performed to determine the connection between operating parameters on chromium removal efficiency (Eq. 3) and the significant parameter was used to predict the final RSM relation equation (Mahmoud and Mahmoud, 2021, Mahmoud et al., 2021a, Maha M. Elshfai and Elsaid, 2021).

$$Y = \beta_0 + \beta_1 x_1 + \beta_2 x_2 + \beta_3 x_3 + \beta_4 x_4 + \beta_5 x_5 \quad (3)$$

where, Y is the predicted percent of chromium adsorption onto GT-nZVI (%); x_1 is pH (2 – 10); x_2 is the adsorbent dose (0.1 – 1.0 g/L); x_3 is stirring rate (100 – 300 rpm); x_4 is contact time (10 – 50min); x_5 is chromium concentrations ranged between (1 to 9 mg/L); β0 is the model intercept; β_1, β_2, β_3, β_4, and β_5 are the linear coefficients of x_1, x_2, x_3, x_4, and x_5, respectively (Rabie S. Farag and Peters, 2019, Mahmoud et al., 2017).

3. Result and dissections

3.1. Characterization

The XRD pattern as shown in Figure (2a) shows two peaks at 2θ = 44.58° and 64.99° indicated the formation of alpha iron structure 110 and 200 planes. The XRD analysis also verified that there are other iron oxides, carbide, iron sodium phosphate, magnesium iron hydride, iron-chromium oxide, iron (III) borate, and ferrimolybdite formations. Figure (2b) shows the SEM image of the prepared GT-nZVI particles before any treatment process. The GT-nZVI showed an irregular surface structure with an average size of about 83 nm. Also, many pores were observed, which help to mass transfer of soluble low molecular weight organic compounds to the inner iron nanoparticles. Figure (2c) shows the EDAX analysis for the selected nanoparticles at 20 mm, indicated the formation of nano-iron particles with outer oxide and carbon and oxygen layers and other elements which come from the extraction solution.

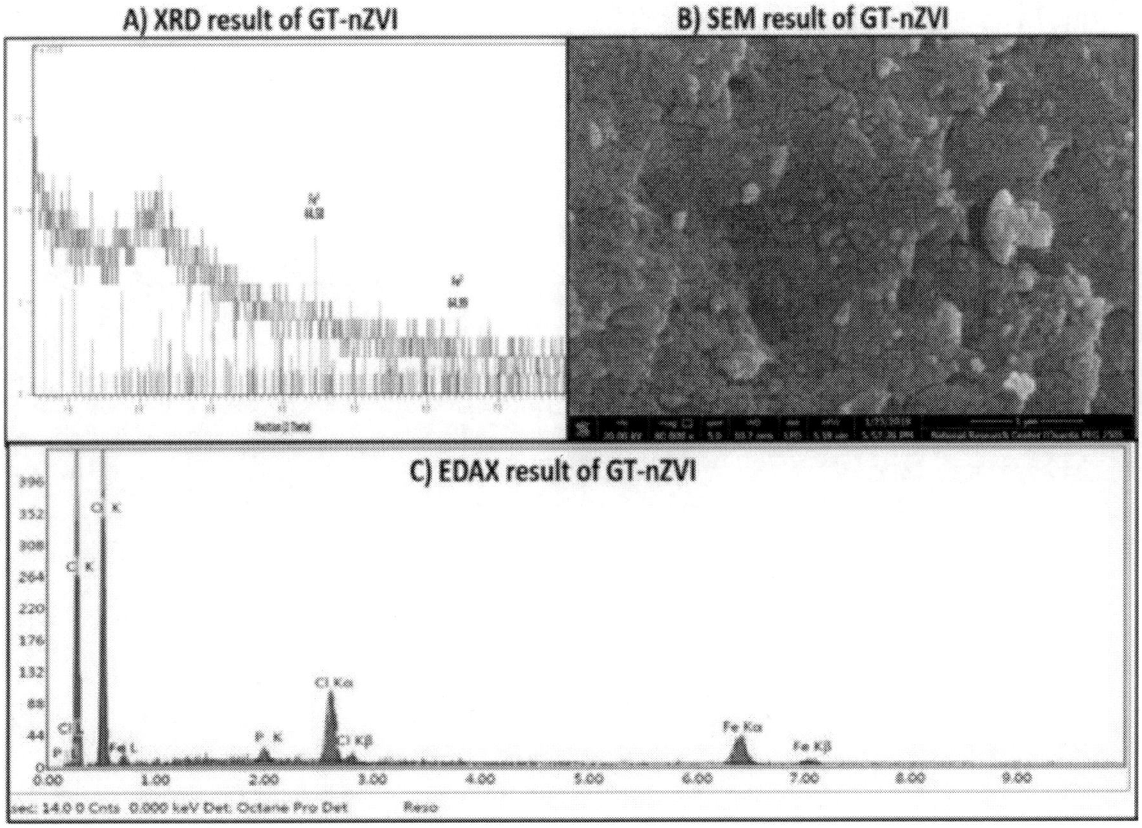

Figure 2. Characterization of GT-nZVI A) XRD of the freshly prepared GT-ZVI, B) SEM of GT-nZVI and C) EDAX results of GT-nZVI.

3.2. Effect of operating parameter

3.2.1. Effect of pH

The pH value is an important parameter affecting chemical and biological reactions in wastewaters (Zhang and Frankenberger, 2003). The obtained results indicated that the most effective pH was 5 showing a removal percentage of 89.6% from the initial chrome concentration of 5 mg/L as shown in Figure 3a. Different studies conducted with the removal of chromium metal ions from aqueous solution using different sorbent materials need high acidic solutions (Gao et al., 2008, Malkoc and Nuhoglu, 2007, Garg et al., 2004, Verma et al., 2006). These phenomena may be explained by that at low pH the surface of rice straw will be surrounded by high quantities of hydronium ions (H^{+}), which will promote the approach of negatively charged $Cr_2O_7{}^{2-}$. The more positive the surface charge of the biomass, the faster the removal rate of Cr (VI) in the rice straw solution, since the binding of anionic Cr (VI) ion species is enhanced. Therefore, solution pH is of great importance for Cr (VI) removal by different sorbent materials (Park et al., 2005).

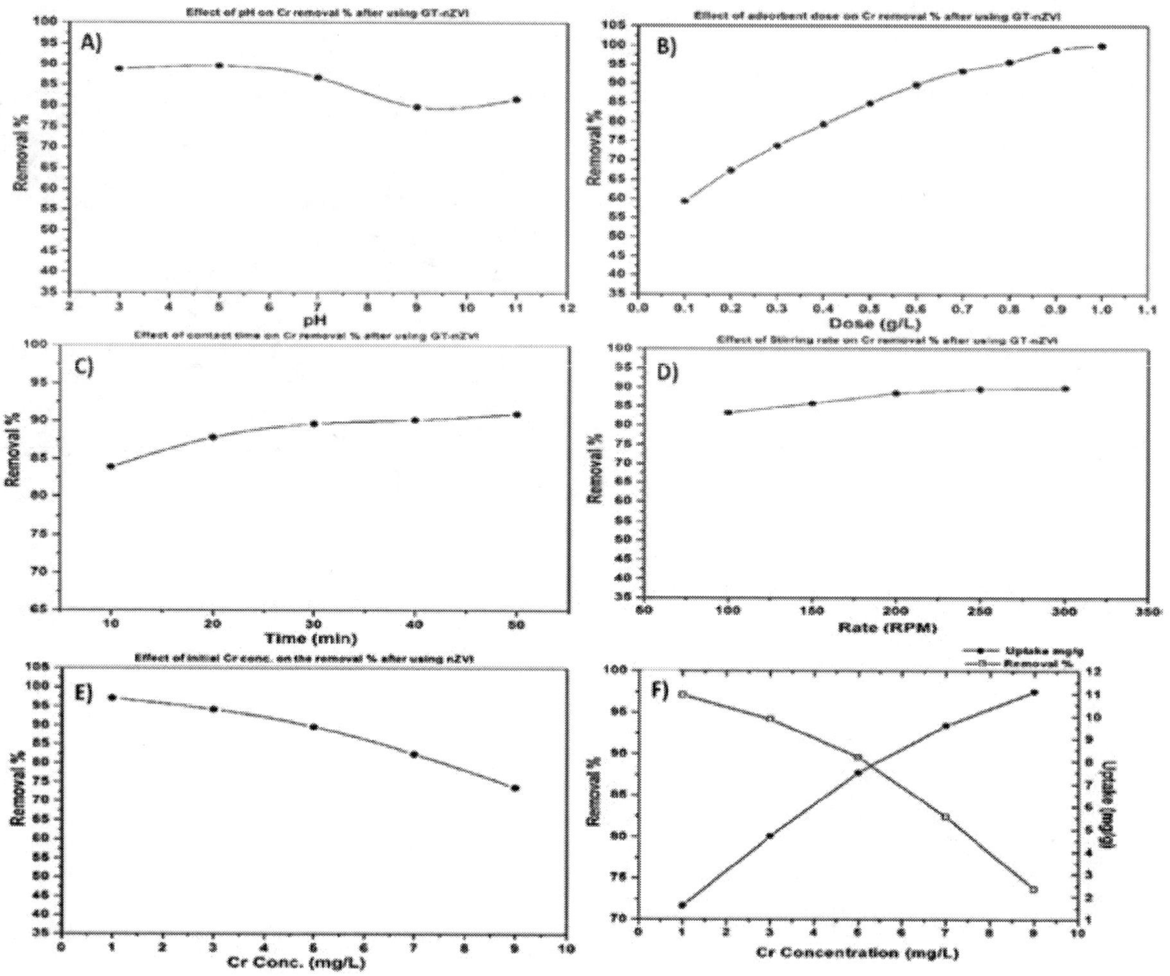

Figure 3. Effect of operating parameter for chromium removal: A) pH effect, B) dose-effect, C) Time effect, D) stirring rate effect, E) concentration effect and F) relation between uptake and removal percent.

3.2.2. Effect of dose

Figure 3b shows the effect of adsorbent dose on chromium removal efficiency after using GT-nZVI as sorbent material. The results showed that the minimum effective adsorbent dose for best removal efficiency is 0.6g/L for showing removal efficiency 89.6. The removal efficiency was increased with dose due to increasing the vacant site for adsorption and free electrons for the degradation Process (Mahmoud, 2017).

3.2.3. Effect of contact time

Figure 3c shows the effect of contact time on chrome removal using at different times namely (10, 20, 30, 40, and 50min). showing that the effective time for best removal efficiency is 30 min. Table 2 explains experimental Qt which is important for kinetic model estimations. Previous studies for Cr removal indicated the effect of contact time for chrome removal using different sorbent materials at pH2, shaking speed 100 rpm and the obtained results indicated that effective time was 10 min (Memon et al., 2009). Another study indicated that the effective time to remove 50 mg/L from chrome solution at pH 2 and stirring rate 250rpm was 60 min (Garg et al., 2009).

Table 2. Effect of contact time on chrome adsorption by using GT-nZVI.

Time (min)	Initial chrome concentration	Final chrome concentration after GT- nZVI	
min	mg/L	mg/L	Qt (mg/g)
10	5	0.805	6.99
20	5	0.61	7.32
30	5	0.52	7.47
40	5	0.495	7.51
50	5	0.455	7.58

3.2.4. Effect of stirring rate

Figure 3d shows the effect of stirring rate on chrome removal using GT-nZVI at different stirring rates namely (100, 150, 200, 250, and 300 rpm). The obtained results indicated that the optimum stirring rate was 250 rpm. Two types of adsorption occur in GT-nZVI surface physical and chemical adsorption. Chemical adsorption depends on the charge between the adsorbed molecule and the sorbent charge. Physical adsorption depends on vacant sites on the surface. Chemical adsorption is more stable than physical adsorption and mainly the stirring rate effect cannot decrease its stability (Ho and McKay, 1998). The amount of adsorbed molecule in physical adsorption equals the summation of adsorbed materials for all individual active sites. The mechanism of the adsorption process depends on the binding energy between adsorbed molecules and sorbents. The adsorption energy decreased gradually until vanishes with the complete adsorption process (Zeldowitsch, 1934). So, the little increase of removal percent according to stirring rate relative to chemisorption reaction and this in agreement with kinetic data (Ho and McKay, 1999).

3.2.5. Effect of initial concentration

Figure 3e shows the effect of initial concentration on chrome removal using GT-nZVI at different concentrations namely (1, 3, 5, 7, and 9 mg/L). The obtained results indicated that the effective chrome concentration was 5 mg/L as shown in Figure 3f.

3.3. Artificial intelligence neural networks (ANNs)

ANNs were trained using the back-propagation algorithm for chrome removal by using GT-nZVI. Network training is a process by which the connection weight and bias on the ANN are adapted through a continuous process of simulation by the environment in which the network is embedded as shown Figure 1 with Rescaling Method for Covariates (Standardized), Activation Function (Hyperbolic tangent) and The Sum of Squares Error were existing in Table 3. GT-nZVI, the sample was trained 15 times and tested 9 times without excluding any result with a total number of 24 runs using network structure 6-4-1. The results indicated that the most significant operating parameter is the effect of concentration 100% followed by the effect of dose 73.1%. The ANN results indicated that there is a small deviation between the predictive and actual results. Also, the residual percent between actual removal % and predicted removal % was between (-6 to +4).

Table 3. Neural network results for chrome removal using GT-nZVI.

ANN model summaries		
Training	Sum of Squares Error	.197
	Relative Error	.028
	Stopping Rule Used	1 consecutive step(s) with no decrease in error
Testing	Sum of Squares Error	1.846
	Relative Error	.329
Independent Variable Importance		
	Importance	**Normalized importance**
pH	.237	65.2%
Dose	.266	73.1%
Stirring rate	.094	25.8%
Contact time	.040	10.9%
Concentration	.364	100.0%

3.4. Response Surface Methodology (RSM)

Table 4 describes the positive linear effect of the all-independent variable "pH", "dose", "time", "stirring rate", and "initial concentration" on chrome removal were observed to be significant ($p< 0.05$). The coefficient of determination between measured data and simulated results (R^2) and adjusted R^2 existed in Table 4. The significant result of the linear RSM model highly significant results agrees with the obtained ANN model to describe the most significant operating parameters for chrome removal using GT- nZVI. By applying Equation 3, the removal equation 4 is obtained. The removal equation can be used to control the removal percent by minimizing or maximizing the operating variables using all obtained results not restricted to the optimum conditions.

$$Y = 77.208 + 1.306\, x_1 + 33.551\, x_2 + 0.027\, x_3 + 0.170\, x_4 - 3.0\, x_5 \quad (4)$$

where, Y is the predicted response of different wastewater contaminants removal efficiency (%); $x_1$is pH (3 – 11); x_2 is adsorbent dose (0.1 – 1.0 g) dry dose and (0.5 – 5 g) wet dose; x_3 is stirring rate (10 – 300 rpm); x_4 is contact time (10 – 50min); x_5 is concentration (1 – 9 mg/L); β_0 is the model intercept; β_1, β_2, β_3, β_4 and β_5 are the linear coefficients of x_1, x_2, x_3, x_4 and x_5 respectively.

Table 4. *t*-statistics and p-values for coefficients of a linear regression model after using GT-nZVI.

		RSM model (Enter Method)					
Model Summary	R	0.970					
	R^2	0.940					
	Adjusted R^2	0.924					
	Std. Error of the Estimate	1.729					
ANOVA	F	56.733					
	Model sig.	00.001					
Coefficients		Term	Estimate	Standard Error	t-Ratio	Prob> |t|	Effect*
	Constant	β0	77.208	3.897	19.812	0.001	Significant
	pH	β1	-1.306	.234	-5.587	0.001	Significant
	Dose	β2	33.551	3.447	9.733	0.001	Significant

Stirring rate	β3	.027	.009	2.877	.010	Significant
Time	β4	.170	.055	3.109	.006	Significant
Concentration	β5	-30.001	.273	-10.97	0.001	Significant

4. Conclusion

Effect of chrome (Cr) removal was studied for initial Cr concentrations (1, 3, 5, 7, and 9 mg/L) and the removal percentages were (97.1, 94.2, 89.6, 82.4, and 73.7%), respectively after using GT-nZVI after applying the optimum conditions at pH 5, dosages 0.6 g, contact time 30 min, and stirring rate 250 rpm. The adsorption isotherms were studied by applying nonlinear equations of Redlich- Peterson, Hill, Sips, Khan, Toth, Koble-Corrigan, Javanovic, Freundlich, and Langmuir models. The obtained results indicated that GT-nZVI preferred Koble–Corrigan isotherm model with the lowest summation of errors 0.394. The kinetic analysis was carried by practicing nonlinear equations of Pseudo First Order (P.F.O), Pseudo Second Order (P.S.O), Elovich, Avrami, and Intraparticle kinetic models. The obtained results indicated that GT-nZVI preferred the Pseudo Second Order mechanism with the lowest summation of errors equal to 0.07. ANN results indicated that GT-nZVI was able to describe the removal of chrome with the maximum residual variations between (-6, +4 %). RSM results indicated that the operating conditions are significant for chrome removal using GT-nZVI.

References

AMERY, H. A. 2002. Water wars in the Middle East: a looming threat. *Geographical Journal,* 168, 313-323.

BAIRD, R. B. 2017. Standard methods for the examination of water and wastewater, 23rd. Water Environment Federation, American Public Health Association, American ....

BARLOW, M. & CLARKE, T. 2017. *Blue gold: the battle against corporate theft of the world's water,* Routledge.

BARNABY, W. 2009. Do nations go to war over water? *Nature,* 458, 282.

BEAUMONT, P. 1994. The myth of water wars and the future of irrigated agriculture in the Middle East. *International journal of water resources development,* 10, 9-21.

CARPENTER, S. R., STANLEY, E. H. & VANDER ZANDEN, M. J. 2011. State of the world's freshwater ecosystems: physical, chemical, and biological changes. *Annual review of Environment and Resources,* 36, 75-99.

CHOWDHURY, M., MOSTAFA, M., BISWAS, T. K. & SAHA, A. K. 2013. Treatment of leather industrial effluents by filtration and coagulation processes. *Water Resources and Industry,* 3, 11-22.

DEL PORTO, D. 2005. Urban and industrial watersheds and ecological sanitation: two sustainable strategies for on-site urban water management. *Water Crisis: Myth or Reality?*, 285.

DIAZ, J. & BARKDOLL, B. Comparison of wastewater treatment in developed and developing countries. World Énvironmental and Water Resource Congress 2006: Examining the Confluence of Environmental and Water Concerns, 2006. 1-10.

DOLATYAR, M. & GRAY, T. S. 2000. The politics of water scarcity in the Middle East. *Environmental Politics,* 9, 65-88.

DUDGEON, D., ARTHINGTON, A. H., GESSNER, M. O., KAWABATA, Z.-I., KNOWLER, D. J., LÉVÊQUE, C., NAIMAN, R. J., PRIEUR-RICHARD, A.-H., SOTO, D. & STIASSNY, M. L. 2006. Freshwater biodiversity: importance, threats, status and conservation challenges. *Biological reviews,* 81, 163-182.

EL-SHAFEI, M., MAHMOUD, A., MOSTAFA, M. & PETERS, R. Effects of entrapped nZVI in alginate polymer on BTEX removal. AIChE Annual Meeting. San Francisco, CA, 2016. 13-18.

ESREY, S. & ANDERSSON, I. Environmental Sanitation from an Eco-Systems Approach. Proceedings of the Workshop on Ecological Sanitation. Mexico, 1999.

FARAG, R. S., ELSHFAI, M. M. & MAHMOUD, A. S. 2018. Adsorption and kinetic studies using nano zero valent iron (nZVI) in the removal of chemical oxygen demand from aqueous solution with response surface methodology and artificial neural network approach. *Journal of Environment & Biotechnology Research,* 7, 12-22.

FORBES, R. J. 1956. *Studies in Ancient Technology Vol V*, Brill Archive.

FROSCH, R. A. & GALLOPOULOS, N. E. 1989. Strategies for manufacturing. *Scientific American,* 261, 144-153.

GAO, H., LIU, Y., ZENG, G., XU, W., LI, T. & XIA, W. 2008. Characterization of Cr (VI) removal from aqueous solutions by a surplus agricultural waste—rice straw. *Journal of Hazardous Materials,* 150, 446-452.

GARG, U. K., KAUR, M., SUD, D. & GARG, V. 2009. Removal of hexavalent chromium from aqueous solution by adsorption on treated sugarcane bagasse using response surface methodological approach. *Desalination,* 249, 475-479.

GARG, V., GUPTA, R., KUMAR, R. & GUPTA, R. 2004. Adsorption of chromium from aqueous solution on treated sawdust. *Bioresource technology,* 92, 79-81.

GUPTA, V. K., GUPTA, M. & SHARMA, S. 2001. Process development for the removal of lead and chromium from aqueous solutions using red mud—an aluminium industry waste. *Water research,* 35, 1125-1134.

HAFEZ, A., EL-MANHARAWY, M. & KHEDR, M. 2002. RO membrane removal of unreacted chromium from spent tanning effluent. A pilot-scale study, Part 2. *Desalination,* 144, 237-242.

HAMDY, A., MOSTAFA, M. K. & NASR, M. 2019. Techno-economic estimation of electroplating wastewater treatment using zero-valent iron nanoparticles: batch optimization, continuous feed, and scaling up studies. *Environmental Science and Pollution Research,* 26, 25372-25385.

HO, Y.-S. & MCKAY, G. 1999. Pseudo-second order model for sorption processes. *Process biochemistry,* 34, 451-465.

HO, Y. & MCKAY, G. 1998. A comparison of chemisorption kinetic models applied to pollutant removal on various sorbents. *Process safety and environmental protection,* 76, 332-340.

ISLAM, M. S. & TANAKA, M. 2004. Impacts of pollution on coastal and marine ecosystems including coastal and marine fisheries and approach for management: a review and synthesis. *Marine pollution bulletin,* 48, 624-649.

JASON A. HEBERLING, Y. G. A., AHMED S. MAHMOUD, MOHAMED K. MOSTAFA, ROBERT W. PETERS. AOP Performance at Wastewater Treatment Plants. ". Proc. 2018 Annual AIChE Meeting, Pittsburgh, PA, (October 28 – November 2). 2018.

KARABAY, S. 2008. *Waste management in leather industry.* DEÜ Fen Bilimleri Enstitüsü.

KENNEDY, P. 2011. *Preparing for the twenty-first century*, Vintage.

KURNIAWAN, A., SISNANDY, V. O. A., TRILESTARI, K., SUNARSO, J., INDRASWATI, N. & ISMADJI, S. 2011. Performance of durian shell waste as high capacity biosorbent for Cr (VI) removal from synthetic wastewater. *Ecological Engineering,* 37, 940-947.

LANGERGRABER, G. & MUELLEGGER, E. 2005. Ecological Sanitation—a way to solve global sanitation problems? *Environment international,* 31, 433-444.

LIU, J. & DIAMOND, J. 2005. China's environment in a globalizing world. *Nature,* 435, 1179.

M ABDELDAYEM, O., ELDAGHAR, O., K MOSTAFA, M., M HABASHY, M., HASSAN, A. A., MAHMOUD, H., M MORSY, K., ABDELRADY, A. & PETERS, R. W. 2020. Mitigation plan and water harvesting of flashflood in arid rural communities using modelling approach: A case study in Afouna village, Egypt. *Water,* 12, 2565.

MAHA M. ELSHFAI, A. S. M. & ELSAID, M. A. Comparative Studies of using nZVI and Entrapped nZVI in Alginate Biopolymer (Ag/ nZVI) for Aqueous Phosphate Removal. 12th international conference on nano-technology for green and sustainable construction, 2021. https://www.researchgate.net/publication …, 21.

MAHMOUD, A. S. 2017. Study the Removal of Most Hazardous Pollutants from Aqueous Solutions Using Nanoparticles. http://lis.cl.cu.edu.eg/search*ara.

MAHMOUD, A. S., FARAG, R. S. & ELSHFAI, M. M. 2020a. Reduction of organic matter from municipal wastewater at low cost using green synthesis nano iron extracted from black tea: Artificial intelligence with regression analysis. *Egyptian Journal of Petroleum,* 29, 9-20.

MAHMOUD, A. S., FARAG, R. S., ELSHFAI, M. M., MOHAMED, L. A. & RAGHEB, S. M. 2019a. Nano zero-valent aluminum (nZVAl) preparation, characterization, and application for the removal of soluble organic matter with artificial intelligence, isotherm study, and kinetic analysis. *Air, Soil and Water Research,* 12, 1178622119878707.

MAHMOUD, A. S., ISMAIL, A., MOSTAFA, M. K., MAHMOUD, M., ALI, W. & SHAWKY, A. M. 2020b. Isotherm and kinetic studies for heptachlor removal from aqueous solution using Fe/Cu nanoparticles, artificial intelligence, and regression analysis. *Separation Science and Technology,* 55, 684-696.

MAHMOUD, A. S., MOHAMED, N. Y., MOSTAFA, M. K. & MAHMOUD, M. S. 2021a. Effective Chromium Adsorption From Aqueous Solutions and Tannery Wastewater Using Bimetallic Fe/Cu Nanoparticles: Response Surface Methodology and Artificial Neural Network. *Air, Soil and Water Research,* 14, 11786221211028162.

MAHMOUD, A. S., MOSTAFA, M. K. & ABDEL-GAWAD, S. A. 2018. Artificial intelligence for the removal of benzene, toluene, ethyl benzene and xylene (BTEX) from aqueous solutions using iron nanoparticles. *Water Supply,* 18, 1650-1663.

MAHMOUD, A. S., MOSTAFA, M. K. & NASR, M. 2019b. Regression model, artificial intelligence, and cost estimation for phosphate adsorption using encapsulated nanoscale zero-valent iron. *Separation Science and Technology,* 54, 13-26.

MAHMOUD, A. S., MOSTAFA, M. K. & PETERS, R. W. 2021b. A prototype of textile wastewater treatment using coagulation and adsorption by Fe/Cu nanoparticles: Techno-economic and scaling-up studies. *Nanomaterials and Nanotechnology,* 11, 18479804211041181.

MAHMOUD, A. S., SARYEL-DEEN, R. A., MOSTAFA, M. K. & PETERS, R. W. Artificial Intelligence for Organochlorine Pesticides Removal from Aqueous Solutions using Entrapped nZVI in Alginate Biopolymer. 2017 Annual AIChE Meeting; Minneapolis, MN, October 29 - November 3, 2017, 2017.

MAHMOUD, M. & MAHMOUD, A. S. 2021. Wastewater treatment using nano bimetallic iron/copper, adsorption isotherm, kinetic studies, and artificial intelligence neural networks. *Emergent Materials*, 1-9.

MALKOC, E. & NUHOGLU, Y. 2007. Potential of tea factory waste for chromium (VI) removal from aqueous solutions: Thermodynamic and kinetic studies. *Separation and purification technology,* 54, 291-298.

MEMON, J. R., MEMON, S. Q., BHANGER, M. I., EL-TURKI, A., HALLAM, K. R. & ALLEN, G. C. 2009. Banana peel: a green and economical sorbent for the selective removal of Cr (VI) from industrial wastewater. *Colloids and surfaces B: Biointerfaces,* 70, 232-237.

MORSY, K. M., MOSTAFA, M. K., ABDALLA, K. Z. & GALAL, M. M. 2020. Life cycle assessment of upgrading primary wastewater treatment plants to secondary treatment including a circular economy approach. *Air, Soil and Water Research,* 13, 1178622120935857.

MOSTAFA, M. & PETERS, R. 2017. Reuse paint wastewater in the manufacture of cement bricks and tiles. *Journal of Material Cycles & Waste Management,* 19.

MOSTAFA, M. & PETERS, R. W. 2016a. A comprehensive assessment of water quality at the Rosetta branch of the Nile River, Egypt. *Journal of Civil Engineering and Architecture,* 10, 513-529.

MOSTAFA, M. K., MAHMOUD, A. S., SARYEL-DEEN, R. A. & PETERS, R. W. Application of Entrapped Nano Zero Valent Iron into Cellulose Acetate Membranes for Domestic Wastewater Treatment. 2017 Annual AIChE Meeting; Minneapolis, MN, October 29 - November 3, 2017, 2017.

MOSTAFA, M. K. & PETERS, R. W. 2015. Use river pollutant modeling to simulate and predict the change in the Damietta Branch water quality before and after construction of the Ethiopian Dam. *Journal of Environmental Protection,* 6, 935.

MOSTAFA, M. K. & PETERS, R. W. 2016b. Improve effluent water quality at Abu-Rawash wastewater treatment plant with the application of coagulants. *Water and Environment Journal,* 30, 88-95.

NELSON, R. R. & ROSENBERG, N. 1993. Technical innovation and national systems. *National innovation systems: A comparative analysis,* 322.

NGOC, U. N. & SCHNITZER, H. 2009. Sustainable solutions for solid waste management in Southeast Asian countries. *Waste management,* 29, 1982-1995.

OKI, T. & KANAE, S. 2006. Global hydrological cycles and world water resources. *science,* 313, 1068-1072.

PAHL-WOSTL, C., MOSTERT, E. & TÀBARA, D. 2008. The growing importance of social learning in water resources management and sustainability science. *Ecology and Society,* 13, 24.

PARK, D., YUN, Y.-S. & PARK, J. M. 2005. Studies on hexavalent chromium biosorption by chemically-treated biomass of Ecklonia sp. *Chemosphere,* 60, 1356-1364.

PIMENTEL, D., BERGER, B., FILIBERTO, D., NEWTON, M., WOLFE, B., KARABINAKIS, E., CLARK, S., POON, E., ABBETT, E. & NANDAGOPAL, S. 2004. Water resources: agricultural and environmental issues. *BioScience,* 54, 909-918.

RABIE S. FARAG, M. M. E.-S., AHMED S. MAHMOUD, MOHAMED K. MOSTAFA, & PETERS, R. W. Green Synthesis of Nano Iron Carbide: Preparation, Characterization and Application for Removal of Phosphate from Aqueous Solutions. Proc. 2018 Annual AIChE Meeting, Pittsburgh, PA, (October 28 – November 2). 2018.

RABIE S. FARAG, M. M. E., AHMED S. MAHMOUD, MOHAMED K. MOSTAFA, AHMED KARAM, & PETERS, R. W. Study the Degradation and Adsorption Processes of Organic Matters from Domestic Wastewater using Chemically Prepared and Green Synthesized Nano Zero-Valent Iron. 2019 Annual AIChE Meeting; Orlando, FL, November 10 - 15, 2019, 2019. https://www.researchgate.net/publication ….

SARYEL-DEEN, R. A., MAHMOUD, A. S., MAHMOUD, M., MOSTAFA, M. K. & PETERS, R. W. Adsorption and Kinetic Studies of using Entrapped Sewage Sludge Ash in the Removal of Chemical Oxygen Demand from Domestic Wastewater, with Artificial Intelligence Approach. 2017 Annual AIChE Meeting, 2017.

SCHWARZENBACH, R. P., EGLI, T., HOFSTETTER, T. B., VON GUNTEN, U. & WEHRLI, B. 2010. Global water pollution and human health. *Annual Review of Environment and Resources,* 35, 109-136.

SHIKLOMANOV, I. A. The world's water resources. International symposium to commemorate the, 1991. 93-105.

SREERAM, K. & RAMASAMI, T. 2003. Sustaining tanning process through conservation, recovery and better utilization of chromium. *Resources, conservation and recycling,* 38, 185-212.

TAYLOR, R. G., SCANLON, B., DÖLL, P., RODELL, M., VAN BEEK, R., WADA, Y., LONGUEVERGNE, L., LEBLANC, M., FAMIGLIETTI, J. S. & EDMUNDS, M. 2013. Ground water and climate change. *Nature climate change,* 3, 322.

VERMA, A., CHAKRABORTY, S. & BASU, J. 2006. Adsorption study of hexavalent chromium using tamarind hull-based adsorbents. *Separation and Purification Technology,* 50, 336-341.

WALSH, A. R. & O'HALLORAN, J. 1996. Chromium speciation in tannery effluent—II. Speciation in the effluent and in a receiving estuary. *Water Research,* 30, 2401-2412.

WEINGÄRTNER, H. & FRANCK, E. U. 2005. Supercritical water as a solvent. *Angewandte Chemie International Edition,* 44, 2672-2692.
WENG, C. N. A critical review of Malaysia's accomplishment on water resources management under agenda 21. Proceedings of International Conference on Environmental Management, Centre for Graduate Studies, National University of Malaysia (UKM), Bangi, 2002. 315-331.
ZELDOWITSCH, J. 1934. Adsorption site energy distribution. *Acta phys. chim. URSS,* 1, 961-973.
ZHANG, Y. & FRANKENBERGER, W. T. 2003. Factors affecting removal of selenate in agricultural drainage water utilizing rice straw. *Science of the total environment,* 305, 207-216.

629405 Enhanced Biodegradability Assessment of Total Petroleum Hydrocarbon By Implementing a Novel Bioaugmentation Strategy of Indigenous Bacterial Consortium

Abstract:

Petroleum industry is the of the main sectors of the world's energy production. A huge amount of petroleum sludge generates during crude oil processing. Being significant in the amount and toxicity of petroleum sludge cause a serious issue to the environment and human health worldwide. The current research intends to isolation and identification of petroleum hydrocarbon degrading bacterial strains from the habitat petroleum refinery sludge. Biodegradation of total petroleum hydrocarbon of petroleum refinery sludge in liquid media by a bacterial consortium having pure culture of four bacterium was performed. The four bacterial strains isolated from the petroleum sludge are *Dietzia* sp. IRB191, *Dietzia* sp. IRB192, *Staphylococcus* sp. BSM19 and *Stenotrophomonas* sp. IRB19. Implementation of novel bioaugmentation strategy was carried out to enhance the biodegradation potential of total petroleum hydrocarbons. Total petroleum hydrocarbon present in the sludge was determined by liquid-liquid solvent extraction. Further, the petroleum fractions were determined by column chromatography. The characterization of oil sludge was performed by ICP-AES, GC-FID, FTIR, SEM analysis. The oil contents asphaltene (50%), and the maltene fraction comprised of 52 ± 3.5% aliphatic, 39 ± 2.5% aromatics, and 9 ± 1% polar fractions, respectively. The ICP-AES analysis depicted that petroleum sludge contains 11 heavy metals (Pb, Cu, Ni, Zn, Fe, Hg, Cd, Mn, As, Cr, and Co) and six metals (Be, Na, Mg, K, Ca, Al). The consortium degrades 77 ± 2.5 %of TPH, whereas the pure culture showed 59-65 % in 2% (w/v) of petroleum sludge in mineral salt media. The degradation of TPH showed a better fit to first order kinetics. This result specified that the developed microbial consortium is suitable for bioremediation process in contaminated oily sludge sites. Considering the high production, toxicity and carcinogenic potential of PRS towards humans, the biotreatment of TPH employing biodegradation is an efficient ecofriendly strategy.

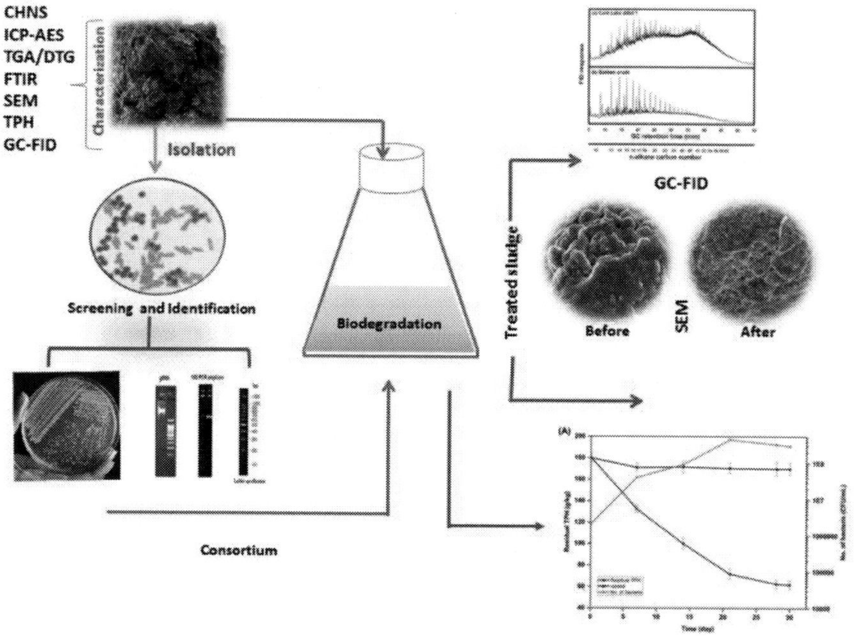

AUTHOR INDEX

AIChE
120 Wall Street, FL 23
New York, NY 10005-4020

ISBN 978-1-7138-5708-2

9 781713 857082

CURRAN ASSOCIATES INC.
proceedings
.com